FARM MACHINES

CULTIVATOR

KATHY MIDDLETON

A Crabtree Roots Book

Crabtree Publishing
crabtreebooks.com

School-to-Home Support for Caregivers and Teachers

This book helps children grow by letting them practice reading. Here are a few guiding questions to help readers build comprehension skills. Possible answers appear here in red.

Before Reading:

- What do I think this book is about?
    - *I think this book is about cultivators.*
    - *I think this book will talk about what cultivators do.*
- What do I want to learn about this topic?
    - *I want to learn about what jobs cultivators do.*
    - *I want to learn why cultivators are important farm machines.*

During Reading:

- I wonder why...
    - *I wonder why farmers use cultivators.*
    - *I wonder why cultivators are needed on the farm.*
- What have I learned so far?
    - *I have learned that cultivators have tines.*
    - *I have learned that cultivators dig and mix soil.*

After Reading:

- What details did I learn about this topic?
    - *I have learned that cultivators get soil ready for planting.*
    - *I learned that cultivators also take out weeds between plants.*
- Read the book again and look for the vocabulary words.
    - *I see the words* ***cultivator*** *on page 3 and the word* ***tines*** *on page 8. The other vocabulary words are found on page 14.*

This is a **cultivator**.

A cultivator digs and mixes **soil** on a farm.

A cultivator attaches to a **tractor**.

Some cultivators have **tines** that dig.

Cultivators get the soil ready for planting seeds.

Cultivators also dig out weeds around **crops**.

Word List

Sight Words

a
also
and
around
attaches
digs
farm
for
get
have
is
mixes
on
out
planting
ready
seeds
some
that
the
this
to
weeds

Words to Know

crops

cultivator

soil

tines

tractor

40 Words

This is a **cultivator**.

A cultivator digs and mixes **soil** on a farm.

A cultivator attaches to a **tractor**.

Some cultivators have **tines** that dig.

Cultivators get the soil ready for planting seeds.

Cultivators also dig out weeds around **crops**.

FARM MACHINES CULTIVATOR

Written by: Kathy Middleton
Designed by: Under the Oaks Media
Proofreader: Crystal Sikkens
Production manager: Candice Campbell
Prepress technician: Katherine Kantor

Photographs: All images from Sutterstock

Crabtree Publishing

crabtreebooks.com 800-387-7650

Printed in the U.S.A./CP2025

Published in Canada
Crabtree Publishing
616 Welland Ave.
St. Catharines, Ontario
L2M 5V6

Published in the United States
Crabtree Publishing
347 Fifth Ave
Suite 1402-145
New York, NY 10016

Library and Archives Canada Cataloguing in Publication
Available at Library and Archives Canada

Library of Congress Cataloging-in-Publication Data
Available at the Library of Congress

Hardcover: 978-1-0398-8105-1
Paperback: 978-1-0398-8466-3
Ebook (pdf): 978-1-0398-8225-6
Epub: 978-1-0398-8345-1
Audio: 978-1-0398-8508-0
Read-Along: 978-1-0398-8550-9